THE POETRY OF POTASSIUM

The Poetry of Potassium

Walter the Educator™

SKB

Silent King Books a WhichHead Imprint

Copyright © 2023 by Walter the Educator™

All rights reserved. No part of this book may be reproduced in any manner whatsoever without written permission except in the case of brief quotations embodied in critical articles and reviews.

First Printing, 2023

Disclaimer
This book is a literary work; poems are not about specific persons, locations, situations, and/or circumstances unless mentioned in a historical context. This book is for entertainment and informational purposes only. The author and publisher offer this information without warranties expressed or implied. No matter the grounds, neither the author nor the publisher will be accountable for any losses, injuries, or other damages caused by the reader's use of this book. The use of this book acknowledges an understanding and acceptance of this disclaimer.

Chemical Element Poetry Book Series
by Walter the Educator™
"Earning a degree in chemistry changed my life!"
– Walter the Educator

dedicated to all the chemistry lovers, like myself, across the world

CONTENTS

Dedication v

Why I Created This Book? 1

One - Essential 2

Two - Masterpiece Of Art 4

Three - Guardian Of Health 6

Four - Potassium, So Pure 8

Five - Enduring Ways 10

Six - Nerves And Muscles 11

Seven - Regulating Balance 13

Eight - Unyielding Role 15

Nine - Sustaining Existence 17

Ten - Regulating Rhythms 19

Eleven - Form And Space 21

Twelve - The Silent Conductor 23

Thirteen - Eternal And Divine	. . . .	25
Fourteen - Potassium, Oh Potassium	. . . . .	27
Fifteen - Nature's Grand Scheme	. . . . . .	29
Sixteen - Battles The Strife	. . . .	31
Seventeen - Richness And Flavor	. . . . . .	32
Eighteen - Every Cell's Story	. . . .	34
Nineteen - Potassium's Presence	. . . . . .	36
Twenty - Across The Land	. . . .	38
Twenty-One - Element So Dear	. . . .	40
Twenty-Two - Chorus Of Living	. . . .	42
Twenty-Three - Mineral Of Abundance	. . .	44
Twenty-Four - Vital Element	. . . .	46
Twenty-Five - We Celebrate You	. . . . . .	48
Twenty-Six - Health And Vitality	. . . . . .	50
Twenty-Seven - Nourishing Our Bodies	. . .	52
Twenty-Eight - Orchestra Of Health	. . . .	54
Twenty-Nine - Day After Day	. . . .	56
Thirty - Crucial Part, Silent, Yet Strong	. . .	57
Thirty-One - Constant Serene	. . . .	59
Thirty-Two - Foods Diverse	. . . .	61

Thirty-Three - Under The Sun's Noon	63
Thirty-Four - Reign Supreme	65
Thirty-Five - Intricate Web Of Life	67
Thirty-Six - Balancing Fluids	69
About The Author	70

WHY I CREATED THIS BOOK?

Creating a poetry book about the chemical element Potassium is a unique and creative way to explore its properties, history, and significance. Poetry has the power to evoke emotions and convey complex ideas in a concise and impactful manner. By using this art form to delve into the world of Potassium, I can shed light on its role in science, industry, and even human health. This approach can make scientific concepts more accessible and engaging to a wider audience, fostering a deeper appreciation for the elements that shape our world.

ONE

ESSENTIAL

In the earth, you lay hidden, yet potent and bright,
A silvery metal, with a secret so light.
Potassium, you dance in the flames of the sun,
Your electrons ablaze, a dance never done.

In the soil and the sea, you quietly reside,
Nourishing life, a humble, unassuming guide.
In bananas and oranges, you show your sweet face,
A mineral friend, in every embrace.

Your atomic number and weight, they define,
A legacy of power, so pure and benign.
From ancient seas, you rose to the land,
A giver of energy, so steady and grand.

In our bodies, you pulse with life-giving force,
Balancing rhythms, staying on course.
With every heartbeat, you whisper your name,
Potassium, essential, in life's vibrant flame.

So here's to you, element nineteen so dear,
Your presence, a comfort, a reason to cheer.
Potassium, oh, how you quietly shine,
A gift from the earth, in every design.

TWO

MASTERPIECE OF ART

In the earth's rich embrace, Potassium lies,
A silent force, hidden from prying eyes.
It whispers through the soil and the sea,
Nourishing life, from root to tree.

In the furnace of the sun, it dances bright,
A fiery performer, a dazzling light.
Its presence in bananas, a cheerful hue,
And oranges too, with their sunny view.

Balancing rhythms, in our bodies it plays,
A conductor of life, in mysterious ways.
From nerve impulses to muscle control,
Potassium's influence, an essential role.

In every heartbeat, a gentle thrum,
Potassium whispers, "I am the drum."
A symphony of cells, in harmonious blend,
Potassium's magic, without an end.

So let's raise a toast to this gift from the earth,
Potassium's quiet shine, of immeasurable worth.
In every design, in every living part,
Potassium hums softly, a masterpiece of art.

THREE

GUARDIAN OF HEALTH

In the orchard's embrace, Potassium lies,
In bananas and oranges, its presence belies.
Balancing rhythms within the beating heart,
Nourishing the body, playing a vital part.

From nerve impulses to muscle control,
Potassium's influence takes its toll.
In the quiet shine of its atomic form,
It weaves through life, a silent storm.

A guardian of health, in every living part,
Potassium dances, a work of art.
Unseen, yet essential, it quietly glows,
In the symphony of life, its presence flows.

So let's raise a toast to this humble element,
For its worth is immeasurable, its role inherent.

In the dance of existence, it takes its place,
Potassium, a hidden force, in every embrace.

FOUR

POTASSIUM, SO PURE

In the earth, you reside, Potassium, so pure,
Nourishing life, an element that will endure.
In soils, you dwell, where roots seek your embrace,
Empowering plants, with your essential grace.

In our bodies, you play a vital role,
Regulating rhythms, from the heart to the soul.
Electrolyte balance, you help to maintain,
Ensuring our well-being, free from any pain.

Potassium, conductor of life's grand symphony,
Guiding the beat, with perfect harmony.
In every cell, your presence is a must,
Sustaining existence, in a continuous thrust.

So here's to you, Potassium, element so grand,
In fruits and veggies, you're found throughout the land.

Balancing rhythms, in nature's endless dance,
An integral part, of life's enchanting trance.

FIVE

ENDURING WAYS

In the earth's embrace, Potassium resides,
A silent guardian, where life abides.
In soil and sea, its presence is found,
Nourishing plants, in cycles profound.

Potassium, the dancer of the body's beat,
Balancing rhythms, in cells so fleet.
Nerves and muscles, it does command,
In harmony with life, its role is grand.

A mineral unseen, yet vital and true,
Guardian of health, in the body's milieu.
Symphony of life, it quietly plays,
Potassium, we toast to your enduring ways.

SIX

NERVES AND MUSCLES

In fields of green, where life does thrive,
Potassium, you keep us alive.
Found in soil, in fruits, and more,
You're essential, that's for sure.

Bananas, oranges, and potatoes too,
Contain the element that's good for you.
Regulating heartbeats, you play a part,
In keeping rhythms, steady from the start.

Inside our cells, you help maintain,
The balance needed for life's refrain.
With nerves and muscles, you have a say,
In how they function, night and day.

So, here's to you, element number nineteen,
In the periodic table, you reign supreme.

Vital for health, in every way,
Potassium, we thank you every day.

SEVEN

REGULATING BALANCE

In the quiet earth, Potassium lies,
A humble guide, beneath the skies.
Nourishing plants with gentle grace,
Empowering life in every place.

In fruits, it hides, a fiery performer,
A quiet shine, a gentle transformer.
Balancing rhythms, in the body's core,
A conductor in life's symphony, forevermore.

Through nerve impulses, it quietly glows,
Controlling muscles, it skillfully shows.
Regulating balance, in cells so grand,
A hidden force, in every living strand.

Oh, Potassium, essential and pure,
In every living part, you endure.

Sustaining existence, with quiet might,
A vital spark, in day and night.
   Grateful we are, for your enduring ways,
For your presence in nights and days.
In every beat, in every breath,
Potassium, we embrace your quiet depth.

EIGHT

UNYIELDING ROLE

In the orchard's embrace, Potassium thrives,
In fruits and greens, where life derives.
Bananas golden, with Potassium's grace,
A mineral that sustains, in every embrace.

Within the body, it plays a vital role,
Balancing rhythms, making the heart whole.
Regulating fluids, it does with ease,
Potassium, the silent guardian, the unseen peace.

Nerve impulses dance, under its command,
Muscles obey, in its steady hand.
Potassium, conductor of life's grand show,
In every movement, its presence does glow.

In the earth's embrace, Potassium resides,
Nurturing plants, where growth abides.
A silent partner in nature's serene dance,
Potassium, the giver of life's chance.

Electrolyte balance, it diligently maintains,
In every cell, its presence sustains.
Vital for life, in every possible way,
Potassium, we honor you, every single day.

So here's to Potassium, in all its enduring ways,
For sustaining life, in its silent blaze.
Grateful we are, for its unyielding role,
Potassium, the essence of life's very soul.

NINE

SUSTAINING EXISTENCE

In the garden of life, Potassium reigns,
Nourishing the earth, it runs through the veins.
From soil to stem, it fosters growth,
In every leaf and fruit, its presence we know.
    Potassium, the silent conductor of life's grand show,
Balancing rhythms, making nature's dance flow.
In every heartbeat, it regulates the pace,
A guardian of health, in its quiet embrace.
    In fruits and veggies, it's a bountiful gift,
A vital nutrient, our spirits it lifts.
From bananas to avocados, it's found in plenty,
A symbol of vitality, in nature's symphony.
    Beyond the orchards, in the human frame,
Potassium plays a crucial role in the game.

Controlling nerve impulses, it keeps us alive,
In harmony with sodium, it helps us thrive.
    So here's to Potassium, in its enduring ways,
Sustaining existence through nights and days.
A humble element, yet mighty and grand,
In life's intricate dance, a partner so grand.

TEN

REGULATING RHYTHMS

In the earth's embrace, you quietly reside,
Potassium, element of grace, so dignified.
Your presence in the soil, a nurturing song,
Enriching the earth, where life belongs.

Within the human frame, you play a vital part,
Regulating rhythms, with a gentle heart.
Controlling nerve impulses, in harmony you sway,
Guiding the dance of life, in your quiet way.

In fruits and greens, your abundance is found,
Nourishing our bodies, with goodness unbound.
Bananas, avocados, and potatoes too,
A testament to the wonders you imbue.

O Potassium, in the symphony of life you sing,
Balancing the stage, with each heartbeat's ring.

From the depths of soil to the beat of our heart,
Your significance endures, a timeless work of art.

ELEVEN

FORM AND SPACE

In the garden's green embrace, Potassium reigns,
Nourishing the earth, it's lifeblood in grains.
From soil to stem, it whispers its might,
Guiding growth with wisdom, day and night.

In cells, it dances, a conductor unseen,
Regulating rhythms, a harmonious sheen.
Nerve impulses obey its gentle command,
In the symphony of life, it takes a stand.

Its presence in fruits, a gift so pure,
Sustaining life, an essence to endure.
Bananas, avocados, a potassium treasure,
A testament to nature's boundless measure.

In the human frame, it plays a crucial role,
Controlling heartbeats, like a guardian soul.
Balancing fluids, maintaining the tide,
Potassium, in its grace, does quietly abide.

So here's to Potassium, steadfast and true,
For nourishing the earth and me and you.
Invisible yet mighty, its presence we embrace,
Potassium, we honor, in every form and space.

TWELVE

THE SILENT CONDUCTOR

In the rhythm of life, Potassium plays its part,
Regulating heartbeats, a conductor at the heart.
Controlling nerve impulses, with a silent command,
Maintaining balance in cells, a maestro so grand.

In the dance of existence, it quietly takes the lead,
Guiding the body, fulfilling every need.
From the earth it rises, in fruits and greens it thrives,
Sustaining life's symphony, where every being strives.

Potassium, oh Potassium, in every living thing,
A quiet force of nature, the melody it does bring.
In the ebb and flow of time, it whispers its decree,
A vital element of life, from sea to shining sea.

So let's raise a toast to Potassium, in all its quiet grace,
For in the grand design of life, it holds a sacred place.

In the orchestra of creation, its presence never wanes, Potassium, the silent conductor, in our veins it remains.

THIRTEEN

ETERNAL AND DIVINE

In the soil deep, your roots entwine,
Potassium, silent strength, divine.
A mineral marvel, hidden from sight,
Nurturing the earth, embracing the light.
In the dance of life, you play your part,
Nourishing plants, with a tender heart.
From grains of sand to fertile ground,
Your presence in soil, profound and unbound.
Through fields and forests, your essence flows,
A symphony of growth, that nature bestows.
In every leaf, in every tree,
Potassium, you sustain life's glee.
In human veins, you course along,
Regulating rhythms, steady and strong.

Nerve impulses hum to your quiet tune,
In harmony with life, from night to noon.
    Oh, essence of vitality, pure and true,
In fruits and veggies, we find you.
Bananas, avocados, and sweet potatoes too,
Potassium, we honor you.
    So here's to Potassium, humble and serene,
In earth and body, you reign supreme.
A silent force, in the grand design,
Potassium, eternal and divine.

FOURTEEN

POTASSIUM, OH POTASSIUM

In the soil, Potassium resides,
Where roots delve deep and life abides.
Nurturing the earth, it plays its part,
Feeding the flora with a gentle heart.

In fruits and veggies, it's found in plenty,
A vital nutrient, so sweet and savory.
Regulating nerve impulses with grace,
In the body's symphony, it finds its place.

From bananas yellow to potatoes brown,
Potassium's presence, a gift profound.
Balancing fluids, it keeps us alive,
In every heartbeat, it helps us thrive.

So let's raise a toast to this element true,
For without it, life would be askew.

Potassium, oh Potassium, we sing to thee,
For sustaining life's grand symphony.

FIFTEEN

NATURE'S GRAND SCHEME

In the earth's embrace, Potassium lies,
A vital element, hidden from prying eyes.
Deep in the soil, it quietly rests,
Nourishing the land, where life divests.

From the roots of plants, it is drawn,
Into the stems and leaves, a bond is spawned.
A catalyst for growth, it plays its part,
In every living organism, it leaves its mark.

In fruits and veggies, it finds its way,
A nutrient essential, come what may.
Bananas, avocados, and sweet potatoes too,
Rich in Potassium, they nourish me and you.

In the human body, it takes the stage,
Regulating heartbeat, and nerve impulses engage.

A partner to sodium, a duo so fine,
Balancing fluids, in this frame of mine.
   So here's to Potassium, in nature's grand scheme,
A silent hero, in soil and bloodstream.
A giver of life, in ways untold,
Potassium, essential, a treasure to behold.

SIXTEEN

BATTLES THE STRIFE

In the earth's embrace, Potassium does dwell,
Nourishing soil where life's stories swell.
A silent hero, in fruits and grains it hides,
A giver of health, where nature abides.

From bananas to spinach, it's nature's gift,
In tomatoes and oranges, its presence lifts.
Balancing fluids, regulating the beat,
Of the heart within, a rhythm so sweet.

In the depths of the earth, its presence is found,
A mineral treasure, in soil it's bound.
A catalyst of growth, a sustainer of life,
In roots and in leaves, it battles the strife.

So here's to Potassium, oh noble element,
In nature's grand design, so truly resplendent.
From the earth to our bodies, its role profound,
A silent hero, in every heartbeat's sound.

SEVENTEEN

RICHNESS AND FLAVOR

In the orchards and fields, you'll find Potassium's might,
A silent guardian, hidden from sight.
Nourishing the earth, it plays a vital role,
In fruits and vegetables, it's the heart and soul.

Bananas, avocados, and sweet potatoes too,
They owe their richness and flavor to Potassium's brew.
Regulating water, conducting nerve signals with ease,
In every living cell, it works to appease.

In soils, it lingers, a giver of health,
Enriching the earth, ensuring abundant wealth.
For plants to thrive, it's an essential part,
Aiding in growth, with its nurturing art.

So let's raise a toast to this element so grand,

For without it, life wouldn't be as grand.
Potassium, oh Potassium, we sing your praise,
For sustaining life in countless ways.

EIGHTEEN

EVERY CELL'S STORY

In the orchard's embrace, Potassium resides,
A silent hero, where nature presides.
Deep in the earth, its roots entwine,
Nourishing life, in a grand design.

In every heartbeat, its rhythm flows,
Regulating balance, as life bestows.
In fruits and veggies, it takes its place,
A vital element, in nature's grace.

Bananas golden, with Potassium's might,
A luscious offering, in morning light.
Spinach and avocados, a verdant song,
Potassium's presence, all along.

From soil to soul, its journey unfolds,
Nurturing growth, as the story beholds.
In cells and tissues, it finds its way,
Sustaining life, night and day.

Oh, Potassium, we sing your praise,
For the symphony of life that you raise.
In nature's palette, you paint your hue,
A vital essence, forever true.

So here's to Potassium, our steadfast friend,
On whom life's dance will always depend.
In every breath, in every cell's story,
Potassium, we hail your silent glory.

NINETEEN

POTASSIUM'S PRESENCE

In the garden of life, Potassium reigns,
Nurturing the earth with its silent grace,
From the roots below to the leaves above,
It feeds the flora with unwavering love.

A catalyst for growth, a giver of life,
In every living organism, it banishes strife,
Regulating nerve impulses with gentle might,
Balancing fluids, keeping everything right.

In fruits and vegetables, it takes its place,
Bestowing health and vigor with its embrace,
A silent hero, in nature's grand design,
Potassium, essential and divine.

So let's raise a toast to this humble element,
For its role in sustaining life's testament,

In every heartbeat and every breath we take,
Potassium's presence, a gift we cannot forsake.

TWENTY

ACROSS THE LAND

In the soil, Potassium lies,
Nurturing the earth, where life thrives.
From the grains to fruits it spreads its love,
Feeding flora, below and above.

Regulating nerve impulses strong,
In every living being, where they belong.
Balancing fluids, a silent chore,
Potassium dances, forever more.

A partner to sodium, in perfect blend,
Balancing fluids, from end to end.
In fruits and veggies, its presence felt,
In every heartbeat, it surely dwelt.

Spinach, avocados, and bananas too,
Potassium's embrace, in every hue.
Nurturing growth, with a gentle hand,
Sustaining life, across the land.

So let's raise a toast to this element grand,
For nurturing life, as it was planned.
In the dance of atoms, it plays its part,
Potassium, the rhythm of the heart.

TWENTY-ONE

ELEMENT SO DEAR

In fields of green, where life takes root,
Potassium whispers, a silent pursuit.
Nurturing growth with a gentle hand,
Balancing fluids, a vital demand.

In fruits and veggies, its presence thrives,
A gift from nature, where abundance thrives.
Bananas, avocados, and sweet potatoes too,
Potassium dances in this colorful view.

Regulating nerve impulses, a symphony profound,
Potassium conducts with a harmonious sound.
From heart to mind, it orchestrates the way,
In every living being, where its influence holds sway.

A mineral gem, in nature's grand scheme,
Potassium shines, like a radiant dream.
In soil and sea, its essence flows,
Sustaining life, where its magic grows.

So here's to Potassium, element so dear,
In every heartbeat, it's quietly near.
A guardian of health, in every form,
Potassium, majestic, through nature's norm.

TWENTY-TWO

CHORUS OF LIVING

In the garden of life, Potassium reigns,
A silent hero, where its presence sustains.
Amidst the earth, in fruits so fair,
It nurtures growth with tender care.

From bananas yellow to potatoes underground,
Potassium's abundance in nature is found.
Regulating nerve impulses with a gentle hand,
Balancing fluids, it helps life expand.

In every cell, its presence is divine,
Conducting vital processes, a symphony so fine.
From heart to muscles, it plays a vital role,
In every living organism, it makes us whole.

Oh Potassium, in the dance of health you partake,
Your importance in life, no one can forsake.
A giver of vigor, a guardian of grace,
In the web of existence, you hold a special place.

So let's raise a toast to this element so grand,
For in the tapestry of life, it takes a stand.
Potassium, we honor your essential might,
In the chorus of living, you shine so bright.

TWENTY-THREE

MINERAL OF ABUNDANCE

In the realm of elements, Potassium reigns supreme,
Balancing fluids, regulating nerve impulses, a vital theme.
From the earth's crust to fruits and greens,
It's a mineral essential, or so it seems.

Potassium, the silent conductor of life's crucial dance,
In every heartbeat, in every chance.
It sparks the neurons, ignites the cells,
In the symphony of existence, its melody swells.

Nature's bounty, in bananas and avocados so fair,
Potassium's presence, beyond compare.
It whispers in the rustle of leaves,
And in the sweetness of fruits, it weaves.

A mineral of abundance, yet often overlooked,

Its importance to life, deeply undershadowed.
But in the grand design of the human frame,
Potassium stands tall, without seeking fame.

So here's to Potassium, the unsung hero of our days,
Invisible yet mighty, in so many ways.
Let's raise a toast to this element divine,
For without it, life's harmony would cease to shine.

TWENTY-FOUR

VITAL ELEMENT

In the dance of atoms, you hold your sway,
Potassium, in your elemental display.
Regulating nerve impulses, you play your part,
Balancing fluids with your chemical art.

In the depths of the earth, you quietly reside,
A vital element, with nature as your guide.
From soil to plant, you work unseen,
Nurturing life in shades of green.

In fruits and vegetables, you abundantly dwell,
Bestowing health with the stories you tell.
Bananas, avocados, and sweet potatoes too,
You're present in foods that nourish and renew.

In the orchestra of life, you take center stage,
Sustaining existence at every age.
Your presence whispers in every living cell,
A testament to the stories you tell.

Oh, Potassium, in your quiet grace,
You weave through the tapestry of every place.
A vital element in the symphony of life,
Invisible yet essential, amidst joy and strife.

TWENTY-FIVE

WE CELEBRATE YOU

In the dance of life, you play your part,
Potassium, element of the heart.
Regulating nerve impulses with grace,
Balancing fluids, you find your place.

In every beat, you keep the rhythm strong,
A silent conductor, where you belong.
Partner to sodium, in perfect blend,
In cells and tissues, your influence extends.

Found in bananas, avocados so green,
In sweet potatoes, a sight to be seen.
In every living organism, you reside,
A humble presence, never to hide.

Nature's gift, sustaining life's flow,
In every breath, in every ebb and flow.
Conducting vital processes, unseen,
In every living being, you reign supreme.

Igniting cells, sparking neurons bright,
In the symphony of life, you shine your light.
Unsung hero, vital and true,
Potassium, we celebrate you.

TWENTY-SIX

HEALTH AND VITALITY

In the garden of life, you quietly reside,
Potassium, a silent conductor, never one to hide.
Regulating nerve impulses, you keep us in tune,
Balancing fluids, under the light of the moon.

Oh, noble K, symbol of strength and grace,
In fruits and veggies, you find your place.
Bananas, avocados, and sweet potatoes too,
Nurturing growth, in all that they do.

Unsung hero, vital to our existence,
Your presence in nature is of great significance.
In the symphony of life, you play a key part,
Quietly conducting, with a steadfast heart.

Though often overlooked, your impact is grand,
Health and vitality, you gently command.

So here's a toast to you, Potassium, so dear,
For without your presence, life would not appear.

TWENTY-SEVEN

NOURISHING OUR BODIES

In the dance of life, Potassium plays its part,
Regulating nerve impulses, a conductor of the heart.
Balancing fluids, it whispers in our veins,
An unsung hero, alleviating life's pains.

In every living cell, it holds its sway,
Guiding vital processes without delay.
A silent guardian, in fruits it hides,
Bananas, avocados, where its presence abides.

From the roots of the earth to the tips of the trees,
Potassium flows, sustaining life with ease.
In the symphony of nature, it plays a vital role,
A humble element, yet it makes us whole.

So here's to Potassium, a treasure unseen,
Nourishing our bodies, keeping us keen.

In the grand tapestry of existence, it weaves,
A quiet force that every living being receives.

TWENTY-EIGHT

ORCHESTRA OF HEALTH

In the garden of life, Potassium reigns supreme,
A conductor in the symphony of existence, a vital theme.
Balancing fluids, it dances with grace,
In every living cell, it finds its place.

From bananas to potatoes, it's nature's embrace,
In oranges and spinach, it leaves its trace.
A silent hero, it keeps the heart beating strong,
In the rhythm of life, it plays along.

Potassium, oh Potassium, we sing to thee,
For without your presence, where would we be?
In the orchestra of health, you take the lead,
A nutrient so vital, in every word and deed.

So let's raise a toast to this element divine,
For in the tapestry of life, it continues to shine.

Potassium, the unsung hero, we honor thee,
For in the grand composition, you hold the key.

TWENTY-NINE

DAY AFTER DAY

In the garden of life, Potassium stands tall,
A silent hero, essential to us all.
From bananas to spinach, it's found everywhere,
Nourishing our bodies with the utmost care.

Potassium, the unsung guardian of our health,
Regulating heartbeat and maintaining our cells' wealth.
A key player in the symphony of existence,
Balancing fluids with quiet persistence.

In every heartbeat, it plays a vital role,
Ensuring our wellness, body, and soul.
From nerve function to muscle control,
Potassium's influence takes its mighty toll.

So here's to Potassium, humble and grand,
A vital element in the intricate band.
For keeping us thriving, day after day,
We thank you, Potassium, in every possible way.

THIRTY

CRUCIAL PART, SILENT, YET STRONG

In the earth's rich embrace, Potassium lies,
A silent conductor, where nature's rhythm lies.
Through fields and orchards, it weaves its gentle thread,
In fruits and veggies, its nurturing touch spread.

In every heartbeat, in every nerve's dance,
Potassium conducts the symphony of life's trance.
Regulating impulses, balancing the flow,
In every living being, its vital presence aglow.

From bananas to avocados, it's a gift of the land,
In sweet potatoes and spinach, its touch so grand.
A guardian of health, a sustainer of might,
Potassium fuels the fire, igniting life's light.

In cells and tissues, it fosters growth's art,
A guardian of vitality, playing its part.

In the grand composition of life's grand design,
Potassium stands tall, a hero so fine.

So let's raise a toast to this element so dear,
For in the dance of existence, its melody's clear.
Potassium, oh Potassium, in nature's grand song,
You play a crucial part, silent, yet strong.

THIRTY-ONE

CONSTANT SERENE

In the garden of life, you quietly stand,
Potassium, the silent hero, so grand.
Regulating rhythms, a conductor unseen,
Balancing fluids, in hues of green.

In every beat of the heart, you play a part,
Nourishing the body, a work of art.
From bananas to avocados, you reside,
In oranges and spinach, you do not hide.

A mineral so vital, you keep us strong,
In fruits and veggies, where you belong.
Nerve impulses dance to your steady tune,
In every cell, beneath the sun and moon.

Oh, Potassium, we sing your praise,
For the role you play in countless ways.
From muscle function to blood pressure's might,
You keep us thriving, day and night.

So here's to you, element number nineteen,
In the table of life, a constant serene.
Potassium, we thank you, for all you do,
In every heartbeat, we feel your value true.

THIRTY-TWO

FOODS DIVERSE

In the earth's embrace, Potassium lies,
A silent guardian in disguise.
Balancing fluids, it does bestow,
In fruits and veggies, its presence does show.

From bananas yellow to spinach green,
In every cell, its presence is keen.
Vital processes, it does aid,
In every life, its role is well-played.

Potassium, oh noble element,
In foods diverse, you are prominent.
Sustaining life with your gentle hand,
In nature's bounty, you take your stand.

From avocados to sweet potatoes,
In grains and legumes, your presence sows.
A vital mineral, you are indeed,
In every meal, you take the lead.

Regulating heartbeat, maintaining cell health,
In every function, you show your stealth.
Potassium, element of great worth,
In nature's abundance, you bring forth mirth.

THIRTY-THREE

UNDER THE SUN'S NOON

In the garden of life, a silent guardian stands,
Potassium, the keeper of our beating hearts.
In every cell, it weaves its magic strands,
Guiding the rhythm of life's intricate parts.

From bananas to spinach, it hides in plain sight,
Nourishing our bodies with its potent might.
Regulating fluids, it holds the key,
To balance and wellness, it sets us free.

Oh, Potassium, you're nature's gift,
A vital force, our spirits you uplift.
In the dance of existence, you play your part,
Sustaining life with your invisible art.

From sunrise to sunset, you silently toil,
In every heartbeat, in every fertile soil.

Weaving through nature's abundant embrace,
Potassium, you're the essence of grace.
    So here's to you, O noble element,
For without your presence, we'd be truly spent.
In the symphony of life, you play your tune,
Potassium, our silent hero, under the sun's noon.

THIRTY-FOUR

REIGN SUPREME

In fields of green, you quietly reside,
Potassium, a force in nature's stride.
Within the earth, your presence lies,
A silent hero, under open skies.

In fruits and veggies, you abundantly dwell,
Nourishing bodies, a story to tell.
Bananas, avocados, and sweet potatoes too,
Your essence brings life, pure and true.

In every heartbeat, you play a part,
Regulating rhythm, a work of art.
Nourishing cells, you keep them sound,
A guardian of health, so profound.

Fluids in balance, thanks to you,
Regulating nerve impulses, pure and true.
A vital force in nature's grand scheme,
Oh, Potassium, you reign supreme.

In sunlit fields and oceans wide,
Your presence sustains, a silent guide.
A mineral so vital, yet often unseen,
Potassium, you keep life serene.

THIRTY-FIVE

INTRICATE WEB OF LIFE

In the garden of life, Potassium reigns,
A silent hero, where health sustains.
Within our cells, it takes its place,
Balancing fluids with silent grace.

From bananas yellow to spinach green,
In fruits and veggies, its presence is seen.
Regulating heartbeats, a vital force,
Potassium keeps life's rhythm on course.

With every pulse, it plays its part,
A mineral keeper of the beating heart.
In nature's grand scheme, it holds the key,
To cell health and vitality, you see.

So let's raise a toast to Potassium's might,
For keeping us strong, day and night.

A mineral so humble, yet so grand,
In the intricate web of life, it stands.

THIRTY-SIX

BALANCING FLUIDS

In the dance of life, you play a crucial part,
Potassium, element of grace and art.
Within our cells, you quietly reside,
Balancing fluids, a gentle guide.

From bananas yellow to potatoes brown,
In fruits and veggies, your presence is found.
Regulating heartbeats, a steady hand,
In nature's bounty, you help us stand.

Silent hero, in every living cell,
You keep us healthy, you serve us well.
A spark of life, in the beating heart,
You play a role, a vital part.

So here's to you, Potassium, we raise our voice,
For your humble presence, we do rejoice.
In the symphony of life, you have a part,
Potassium, element of grace and art.

ABOUT THE AUTHOR

Walter the Educator is one of the pseudonyms for Walter Anderson. Formally educated in Chemistry, Business, and Education, he is an educator, an author, a diverse entrepreneur, and he is the son of a disabled war veteran. "Walter the Educator" shares his time between educating and creating. He holds interests and owns several creative projects that entertain, enlighten, enhance, and educate, hoping to inspire and motivate you.

Follow, find new works, and stay up to date
with Walter the Educator™
at WaltertheEducator.com

www.ingramcontent.com/pod-product-compliance
Lightning Source LLC
LaVergne TN
LVHW052001060526
838201LV00059B/3766